AF270579

SCIENCE QUESTIONS

# HOW DOES PLUMBING WORK?

by Elizabeth Andrews

**Cody Koala**

An Imprint of Pop!
popbooksonline.com

abdobooks.com

Published by Pop!, a division of ABDO, PO Box 398166, Minneapolis, Minnesota 55439. Copyright ©2022 by Abdo Consulting Group, Inc. International copyrights reserved in all countries. No part of this book may be reproduced in any form without written permission from the publisher. Cody Koala™ is a trademark and logo of Pop!.

Printed in the United States of America, North Mankato, Minnesota

102021
012022

THIS BOOK CONTAINS RECYCLED MATERIALS

Cover Photo: Shutterstock Images
Interior Photos: Getty Images, 6, 9, 13 (top), 16; Shutterstock Images, 5, 7, 10, 13 (bottom row), 14, 15, 17, 19, 20

Editor: Grace Hansen
Series Designer: Laura Graphenteen

Library of Congress Control Number: 2021942299
Publisher's Cataloging-in-Publication Data
Names: Andrews, Elizabeth, author.
Title: How does plumbing work? / by Elizabeth Andrews
Description: Minneapolis, Minnesota : Pop!, 2022 | Series: Science questions | Includes online resources and index.
Identifiers: ISBN 9781098241100 (lib. bdg.) | ISBN 9781098241803 (ebook)
Subjects: LCSH: Plumbing--Juvenile literature. | Water-pipes--Juvenile literature. | Drainage pipes--Juvenile literature. | Children's questions and answers--Juvenile literature.
Classification: DDC 500--dc23

## Hello! My name is

# Cody Koala

Pop open this book and you'll find QR codes like this one, loaded with information, so you can learn even more!

Scan this code* and others like it while you read, or visit the website below to make this book pop.

**popbooksonline.com/plumbing**

*Scanning QR codes requires a web-enabled smart device with a QR code reader app and a camera.

# Table of Contents

# Chapter 1

# What Is Plumbing?

Is it magic that makes water fall from a faucet at the twist of a knob? It must be! But it's not magic. It's just plumbing!

Watch a video here!

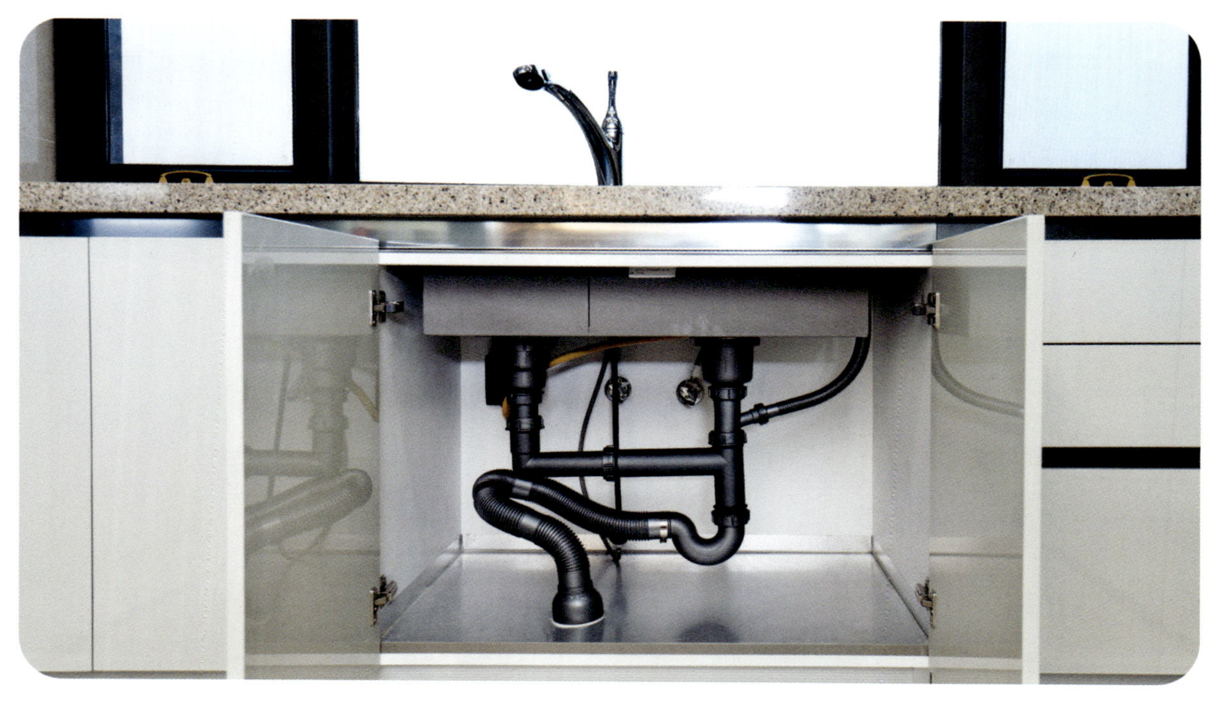

Plumbing has to do with
the pipes that bring clean,
safe water to homes.

Plumbing also moves waste away from the home. A person who fixes or places plumbing is called a plumber.

# Chapter 2

# Why Is Plumbing Important?

People need plumbing to stay healthy. In the past people would dump waste into water sources. **Diseases** can grow in unclean water. Many people got sick and died when they used the water.

WARNING
POLLUTED
WATER
Unsafe for Drinking
or Recreational Use
DIRECTOR OF PUBLIC HEALTH
DEPARTMENT OF PUBLIC HEALTH
Learn more here!

In the 1800s people began to understand the importance of clean water. Cities thought about ways to safely move waste out of homes. They also made better **systems** to get clean water into homes.

# The Two Types

Plumbing today is put together with metal or plastic materials. The water is moved with high-**pressure pumps**.

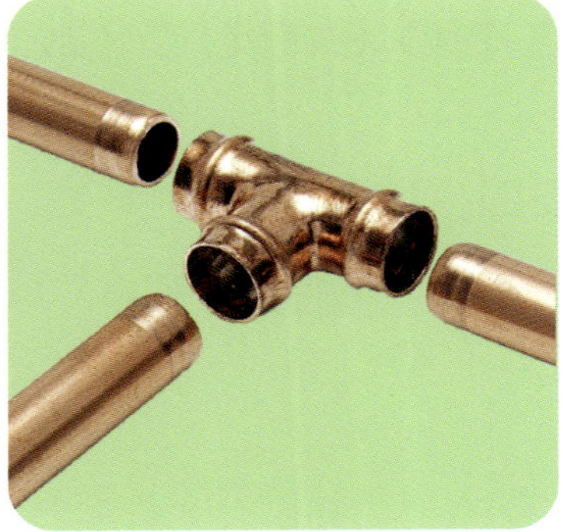

Learn more here!

There are two ways to get water to your home. The first way to get water is from a well. A well is a hole dug deep into the ground. It reaches stored water in an **aquifer**.

The water is pumped up into pipes called water mains. These lead to a storage tank. The tank is usually in the basement of a home.

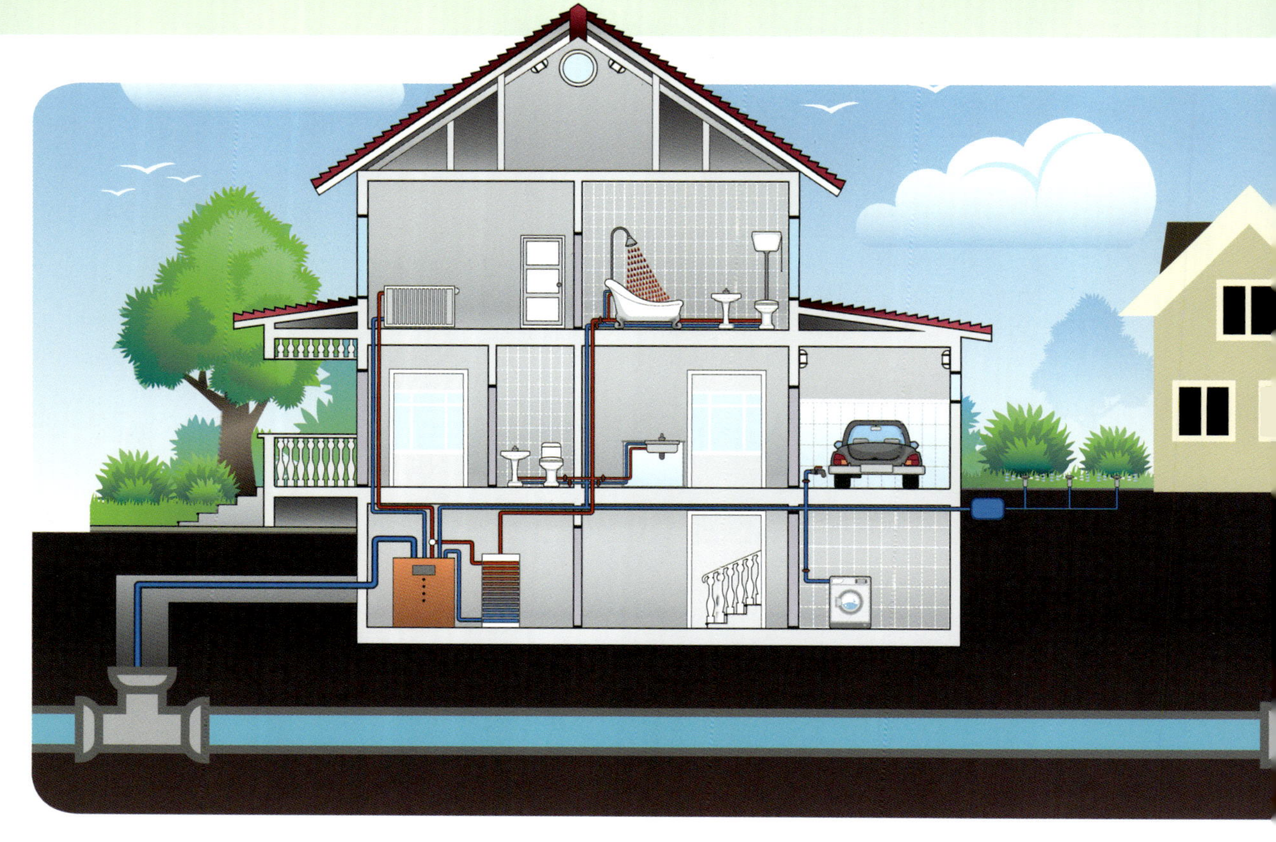

Another way to get water is from the city. Cities store water from lakes and rivers. The water could also come from aquifers.

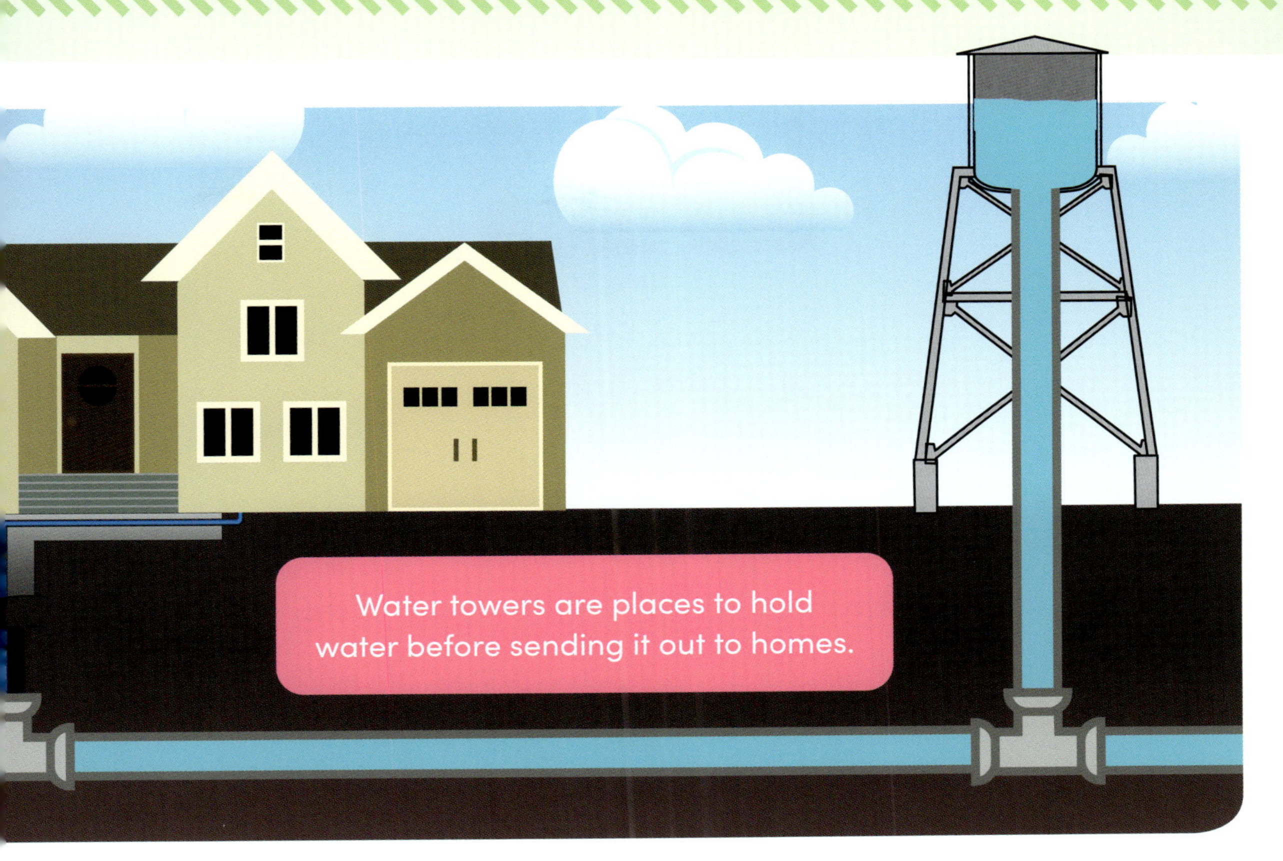



It is pumped into a water main that goes around the city. Then it is pulled into the home through pipes called lateral lines.

# Chapter 4

# Safe Water

All water that goes to people's homes must be filtered and tested. This keeps water safe to drink and bathe with. Safe water is called potable water. The water that goes down the drain is called wastewater.

Complete an
activity here!

Cities and people have to be smart about how they use water. Sometimes an area might go through a **drought**. There will be less water available. To prepare for this, cities store large amounts of water. This is called a reservoir.

# Making Connections

## Text-to-Self

Were you curious about how plumbing worked? Are there any other things in your home you wished you knew more about?

## Text-to-Text

Droughts are mentioned in this book. Have you read any other books that talk about droughts?

## Text-to-World

What do you think the world was like before people had plumbing?

# Glossary

**aquifer** – a layer of rock, sand, or gravel that can absorb water.

**disease** – a sickness.

**drought** – a long period of dry weather.

**pressure** – the pushing of a force against an opposing force.

**pump** – to raise or move using a pump. A pump is a tool that raises or moves liquid or gas into or out of something.

**system** – a set of things working together.

# Index